The Marshall Cavendish Library of Science Projects

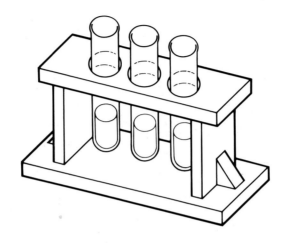

Marshall Cavendish Corporation

London • New York • Toronto

The Marshall Cavendish
SCIENCE PROJECT BOOK
of
WATER

Written by Steve Parker
Illustrated by David Parr

Reference Edition published 1989

The Marshall Cavendish Library of Science projects
Water Volume 1

© Marshall Cavendish Limited MCMLXXXVI
© Templar Publishing Limited MCMLXXXV
Illustrations © Templar Publishing Limited MCMLXXXV

Trade edition published by Granada Publishing Limited

Reference edition published by:
  Marshall Cavendish Corporation
  147 West Merrick Road
  Freeport
  Long Island
  NY 11520

Printed and bound in Italy
by L.E.G.O. s.p.a., Vicenza

Library of Congress Cataloging in Publication Data

The Marshall Cavendish Library of Science Projects

  Includes index.
  Contents: (1) Water
  1. Science – Experiments – Juvenile Literature.
2. Science – Juvenile Literature. (1. Science –
Experiments. 2. Experiments) 1. Marshall Cavendish
Corporation.
Q164.M28   1986   507.8   86-11731
ISBN 0-86307-624-6 (Set)

ISBN 0-86307-627-0 (Vol 1 Water)

CONTENTS

Golden rules 4
Waterworld! 6
What's in water? 8
Experiments 10
We all need water 12
Water at work 14
Experiments 16
Water in disguise 18
Ice and steam 20
Experiments 28
Wet or dry? 24
Endless journeys 26
Experiments 28
Sink or swim? 30
Floating about 32
Experiments 34
Walking on water 36
Wetting & washing 38
Experiments 40
Things to remember 42

Science is all about discovering more about your world, finding out why certain things happen and how we can use them to help us in our everyday lives. SCIENCE PROJECTS looks at all these things. It's packed with exciting experiments and projects for you to do, and fascinating facts for you to remember. It will teach you more about the world around you and to understand how it works.

GOLDEN RULES

This book contains lots of scientific facts and experiments to help you find out more about water and its strange properties. Whenever you try one of the experiments, make sure you read all about it before you start. You'll find a list of all the things you need, a step-by-step account of what to do, and finally an explanation of why and how your experiment works.

▶ Always watch what happens very carefully when you're doing an experiment and, if you find it doesn't work the first time, *don't* give up. Consider what could have gone wrong. Are the conditions right? Perhaps the temperature of the room or a draft could have affected your results. Read through the experiment once more, check that everything is just right, and then try, try again. Real scientists may have to do an experiment several times before getting a worthwhile result.

▶ Because you will be such an active scientist, it's a good idea to start collecting for your laboratory. Nearly everything you need for the experiments can be found around your home. For example, bottles and jars will often be used, so when you

▰▰GOOD SCIENTISTS...▰▰

ALWAYS THINK SAFETY FIRST

Famous scientists take precautions to avoid danger, so that they live to see their results and enjoy their fame. In any project or experiment, especially one you have thought up yourself, consider what it is you are trying to prove and have a good idea of what should happen. Don't do any experiment without planning it 'just to see what happens'.

ALWAYS KEEP A NOTEBOOK

Whenever you are involved in scientific activity, keep your *Science Notebook* by your side and fill it with notes and sketches as you go along. Get into the habit of writing up your experiments and observations – your notes will come in handy in the future.

ALWAYS FOLLOW GOOD ADVICE

Advice and instructions, like the leaflets that come with pieces of equipment, should be read and understood. They are there for your safety and help. Good scientists think for themselves, but they are also wise and listen to what others have to say.

see your parents throwing away useful containers offer to wash them, and then add them to your collection. Collect various chemical substances – washing soda, salt, sugar, lemon juice and dishwashing liquid, for example – and put them in labeled jars in a safe place. General things like pens, paper, a ruler, spoons for measuring and a pair of scissors will also come in handy. You'll need a work surface for your experiments and it's a good idea for this to be near a sink. Store your materials in a nearby cupboard or cardboard box.

► Always let your parents know what you are doing. Sometimes you'll need their help. And when it comes to special equipment like light bulbs or batteries, they'll know where to get it. Your parents may help you to build wooden stands or nail things down when needed. And if you need to use matches or cut things out, ask their permission first.

► Good scientists always clean up when they have finished! After you have done your experiments, throw away anything you won't need again and clean everything else, ready for the next experiment.

NEVER MESS WITH THE MAIN LINE

Don't play with main line electricity or electrical outlets. Carry out all your electrical experiments with low-power batteries (preferably 3 or 4½ volts). Remember, main line electricity can kill you.

NEVER PLAY WITH CHEMICALS

Avoid mixing chemicals and powders unless you are sure that you know what is going to happen, and always use small quantities. Dangerous chemical mixtures can explode or start a fire or burn your eyes and skin. Make sure any chemicals you keep are properly stored in jars and are correctly labeled.

NEVER FOOL WITH HIGH-PRESSURE EQUIPMENT

Do not play around with gas or liquids under pressure, especially in containers like aerosol cans – even if they seem empty. They can blow up in your face. Dispose of empty aerosols carefully and *never* put them in or near the fire.

WATERWORLD!

magine you are a space traveler on your first visit to the planet Earth. As you go into orbit before landing, you see that over seven-tenths of the Earth is not land. Instead, it's a vast, flat sheet of a peculiar blue-green substance.

Upon landing the Earth beings seem friendly enough, and they offer you some refreshment. You are very cautious about what they might serve so you ask for their most harmless and basic substance. An Earthling brings you what looks like an empty glass. In the glass is a substance that has no color, no taste, and no smell — but it is quite heavy and seems to splash around and spill if you aren't careful. You test the substance to check it is safe. Your powerful hand-held scientific analyzer shows it to be a liquid, made up of billions of incredibly small things called molecules. All the molecules are exactly the same. Each one is made up of three parts, called atoms — two are atoms of hydrogen and the other is an atom of oxygen (written H_2O). Beyond this your

WHAT IS WATER MADE OF?

Water is made of molecules. Each one has two atoms of hydrogen and one of oxygen.

Molecule

Oxygen atom

Hydrogen atoms

Photo: Science Photo Library / A Hart-Davis

analyzer shows no readings.

After a few days on Earth you discover that this substance is absolutely everywhere. It falls from the sky in tiny droplets called rain. It flows along channels in the ground called rivers, and into large holes called lakes. Factories use huge amounts in their industrial processes, and in practically every dwelling place there are tubes with handles that stick out from the wall and when you turn the handle out comes this stuff! The Earthlings drink it, cook in it, wash themselves in it, clean their clothes and vehicles in it, and even spray it on their plants. In fact, nine-tenths of each plant is made up of this substance, and even three-quarters of each Earth being!

When you finally learn the name of this universal substance you might think that perhaps the inhabitants have mis-named their world. It shouldn't be called Earth it should be called the planet Water...

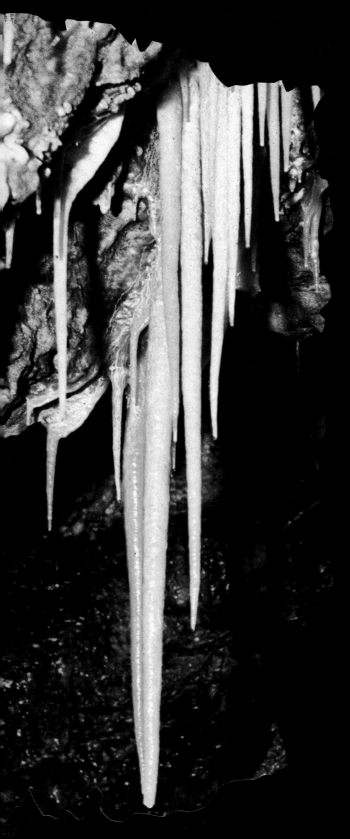

WHAT'S IN WATER?

Take a look at some water. Although it might appear to be pretty clean and clear, it's a fact that most water has got all sorts of other things mixed up in it, quite apart from its atoms of hydrogen and oxygen. You probably know that sea water contains salt, for instance, but did you realize that even the water from your tap at home isn't absolutely pure?

What water leaves behind

Have you ever been into a cave and seen those marvelous stone columns called stalactites and stalagmites? You probably know that these great, solid columns are, in fact, growing slowly – by about $1/100$ inch ($1/4$ mm) every year. But the scientist's inquiring mind asks: where does the new stone come from, that adds to the growing column? The only answer seems to be the clear, pure water that runs over the stone and drip-drips onto the cold rock below. So is the water really clean and pure? Well, the answer is, of course, 'no'. Although it looks clear, the water actually contains tiny amounts of various minerals that have become added to it on its journey from the sky through the earth. The process starts with the rain water that trickles through the soil and *dissolves* various minerals to form a *solution*. As the solution drips into the cave and over the rock, some of the water *evaporates* and leaves behind its extra load of minerals. So it's the minerals that form the small amounts of solid stone that gradually add to the lengthening columns.

Putting water to work

Stalactites and stalagmites are a good example of how dissolving and evaporation happen quite naturally. But very often

Science factfile

Growing underground

Look in an underground cave or cavern and you'll probably find some stalactites and stalagmites growing there. Some hang from the ceilings, like those in the picture on the left, while others rise up from the floors, sometimes as solid 'drips', sometimes as slender pillars.

► The biggest stalagmite in the world can be found in the Aven Armand cave in Lozère, France. It's about 97 feet (29 meters) high – that's about four times taller than the average house – and it is still growing!

► The longest free-hanging stalactite has so far reached a length of 23 feet (7 meters) in the Poll – an ancient cave in County Clare, Ireland.

► Some stalactites and stalagmites have become so famous they've been given special name. The Bicentennial Column in Ogle Cave, New Mexico, USA is one example.

► Scientists discovered that these strange rock formations grow about $\frac{1}{100}$ inch ($\frac{1}{4}$mm) each year. Can you work out how long it has taken these record-holding stalactites and stalagmites to grow this big?

Science project

Now you see it; now you don't!

You can see how water dissolves small amounts of other substances and then leaves them behind again, by following the steps below.

1 The idea of *dissolving* is easy to demonstrate. Put a spoonful of salt into a glass of tap water and stir. The salt seems to disappear – but it's still there. You can tell this by dipping your finger in and then tasting it. Yuk!

2 What you have is a *solution* of salt in water. Salt is called the *solute*, and water is the *solvent*. So next time you're asked "Do you take sugar in your tea?", you can reply "Yes please, I'll have one spoonful of solute in one cup of solvent"!

SOLUTE   SOLVENT

3 *Evaporation* works the other way round. Put the salty water in a pan and leave it in a warm place for a few days. The water will evaporate to leave your spoonful of salt caked on the bottom of the pan!

they are made to happen by us, to help in our everyday lives. Some examples are listed here; can you think of any others?

► Sea salt for seasoning our food is made by flooding large pools with sea water, and then letting the hot sun evaporate the water to leave behind the solid salt.

► Colored dyes are pigment dissolved in water to make a solution. A garment is soaked in the solution and then dried. The water evaporates to leave behind the pigment which colors the garment.

► If you make up some water-based paint like water-color, you are making a solution of the pigment in water. The when you paint a picture, the paint is brushed onto the paper and the water evaporates to leave behind the dry pigment.

here are two experiments to help you prove for yourself some of water's strange properties. Find out how to make ordinary tap water release a few of its atoms of oxygen and hydrogen for you to see. And discover how to make your own stalagmites and stalactites without going underground!

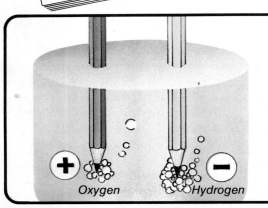

Experiment

Splitting up water

In this simple experiment you can use a small electric current to split up water molecules so they release a few of their hydrogen and oxygen atoms as gas. You'll need a glass jam jar full of tap water, two lead pencils sharpened at both ends, a 9 volt battery, some 15 amp fuse wire, a pair of scissors, a square of cardboard and some sticky tape.

step 1

Place your square of cardboard over the jam jar and push the pencils through so their points are held in the water.

step 2

Cut two pieces of fuse wire about 10 in (25 cms) long. Wind one end of each piece around the battery terminals and fix them in place with sticky tape. Make loops at the other ends and stick them round the lead points of each pencil.

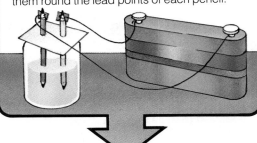

step 3

When the wires touch the pencils, the electric current from the battery will flow down one lead, through the water and back up the other lead. You will soon see tiny bubbles forming around the submerged pencil points. The bubbles around the pencil attached to the positive battery terminal (marked +) are bubbles of oxygen gas. And those around the negative one (marked −) are bubbles of hydrogen. This process is known as *electrolysis*. You will notice more hydrogen bubbles than oxygen. Do you know why?

Oxygen Hydrogen

Experiment 2

Making crystal columns

You can make your own stalactites and stalagmites just like those on page 8 by following this simple experiment. All you need are two glass jam jars, a plate and a spoon, some baking soda, some pieces of wool twisted together to make one thick strand, and some water.

Step 1

Fill your jars with hot water from the tap. Stir in lots of baking soda until no more will dissolve. If you add any more it will fall to the bottom because the water has absorbed as much soda as it can hold. When this happens, you'll have made what is called a *saturated* solution.

Step 2

Put the jars in a warm place and position the plate in between them. Drop the ends of the wool into the jars so that the middle hangs over the plate. Gradually the two soda solutions will creep along the woolly strand until they meet in the middle. The solution will then start to drop onto the plate.

Step 3

After several days you will see mini-stalactites and stalagmites beginning to form on the wool and the plate. This happens because the water in the tiny drops of solution evaporates as it drips from the wool. The water turns into vapor, leaving behind the soda particles as a hard deposit. Look at your 'stalactite' with a magnifying glass. You'll see it is made up of thousands of soda crystals.

If you leave your jars for a few more days your columns may eventually meet in the middle. They will have formed in much the same way as the stone stalactites and stalagmites only much, much quicker. How long do you think it would take to make your soda columns as big as the record-breakers mentioned on page 9?

WE ALL NEED WATER

A person can go without food for several weeks if he really has to, but he would only survive a few days without water. Look at a map and you'll see that most big towns and cities have rivers flowing through them. This is no coincidence – the original reason for people settling there was the supply of water.

Of course, it's not only humans that need fresh water. Ask a zoo-keeper what his animals need most and he'll probably say: a bowl of fresh water every day. Ask a farmer if it's going to be a good harvest and he may reply: yes, as long as we have plenty of good spring rain to make the seeds grow. All living things, from the tiniest insect to the tallest tree, need a supply of water to live. And, what's more, they are all made partly of it. Even you are three-quarters water. And a lettuce leaf is nineteen-twentieths water! So what makes water so vital?

The answer is that water is an excellent solvent (as we saw on page 9). The chemical processes that are necessary for things to live can only happen in solution. Every living thing is made up of billions and billions of molecules, and it is only when the molecules are dissolved in water that they can move about freely and combine with each other in various chemical reactions.

All living things need water, but some require more than others. A fish needs more than a camel, for instance, but even a camel has to drink. A camel can last for three weeks without water, but when it does have a drink it takes in over 30 gallons (about 120 liters) (more than a

TOWNS AND RIVERS

Look at a map of your area. Are there any rivers nearby? The chances are that there will be at least one. Notice how many towns and villages are built near the water. Can you find any evidence of water being used by the people in the towns – a water mill would be a good example?

bathtubfull)! A cactus plant 'drinks' in the same way. When the rains come, it absorbs as much water through its special roots as its spongy stem can take, and becomes over nine-tenths water, Then it uses this reservoir very slowly.

Just think for a moment of all the

Water makes the desert bloom, as you can see by the trees growing around this oasis. Just a few yards away the desert is parched and lifeless.

HOW MUCH WATER?

Human 75%

Jellyfish 95%

Cucumber 95%

Wood 10%

Egg 74%

ways your body uses water to help you survive and live more comfortably. Your eyes water to cleanse them of any dust or dirt, you sweat salty water to help you cool down when you're hot, and you wash away all your body wastes in water every single day!

Plants use water too – they absorb it from the soil and use it to make food so they can grow new leaves, shoots and flowers. You can find out more about plants and water on page 16.

13

WATER AT WORK

You probably don't think much about the water in your home. You just turn on the tap and out it comes. But there is a whole industry organized to supply your home with fresh, clean water, and when something goes wrong everybody not[ ]. It might be a dripping tap, or a burst pipe, or a leak in the central heating – or even a drought, when main line water is cut off and you have to carry all your supplies from a standpipe in the road. Only then do you realize how important water is, and how much you use every day.

Using water

You might think that lots of clean water is absolutely essential. We drink in it, cook in it, wash in it and flush the toilet with it. We clean cups, plates, pots, pans, clothes and cars in it. In an average day you might use 50 gallons (around 200 liters) of water on all these activities. But out in the desert, a nomad can live with less than a gallon (1 or 2 liters) of water a day, for drinking, a bit of cooking and the occasional wash.

Industry uses water hundreds and thousands of times faster than we do at home. To make just one car takes about 4,000 gallons (around 15,000 liters). A modern factory would be paralyzed without water.

Giant power stations also use water, to cool their turbines. You can sometimes see the hot water steaming from the huge cooling towers. This is a much bigger version of the radiator in a car, which uses water to carry heat away from the car engine so that it does not get too hot.

Powerful water

Water is powerful, too. This is because it is heavy and flows easily. Running water has been used for centuries as a source

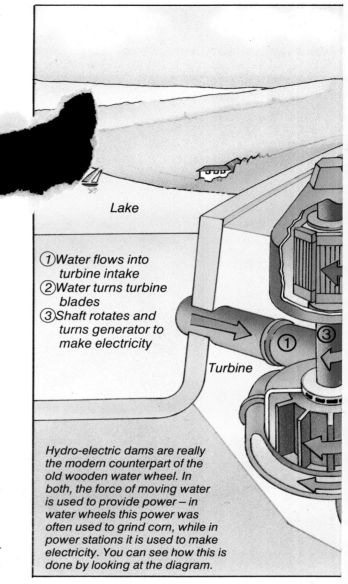

Lake

① Water flows into turbine intake
② Water turns turbine blades
③ Shaft rotates and turns generator to make electricity

Turbine

Hydro-electric dams are really the modern counterpart of the old wooden water wheel. In both, the force of moving water is used to provide power – in water wheels this power was often used to grind corn, while in power stations it is used to make electricity. You can see how this is done by looking at the diagram.

Science factfile

Water power

► Water mills have been around for a long, long time. The oldest one in Britain was used as a corn mill and can be found at Priston Mill near Bath. It has been on record since the year 931, which make it over one thousand years old!

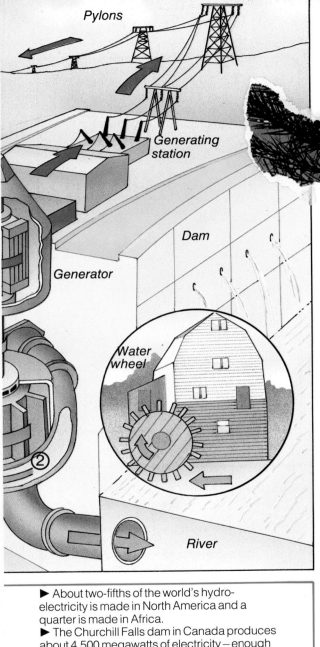

Pylons

Generating station

Dam

Generator

Water wheel

②

River

of energy. The energy comes from water 'falling' as it flows along a river or over a waterfall. Watermills were built on the banks of fast streams, and the power was used to turn stones and grind wheat into flour, to make bread. Modern hydro-electric dams channel water over large turbines that turn generators to produce ~~ctricity~~. Once the dam is built, there is ~~little~~ in the way of running costs and ~~hydro~~-electricity is an important source ~~energy~~ for the future. It has its ~~drawbacks~~, though. Sometimes whole valleys are flooded to capture enough water to make the dam worthwhile (though the water is often used to irrigate nearby fields). Niagara Falls, on the American/Canadian border, has less than half the water cascading over it that it used to – the other half has been diverted to produce electricity.

► About two-fifths of the world's hydro-electricity is made in North America and a quarter is made in Africa.
► The Churchill Falls dam in Canada produces about 4,500 megawatts of electricity – enough to light 45 million light bulbs or run 80 million home computers.
► The world's first tidal power station was opened just off the coast of Brittany, France, in 1966. It cost millions of dollars to build and harnesses so much power from the sea that it has imperceptibly slowed the Earth's rate of revolution!

Science in action

Out of the tap and back again

How many different processes do you think water has to go through before it is ready for you to drink? Well, the answer is loads! See if you can find out the sites of local reservoirs and pumping stations and visit them with a grown-up.

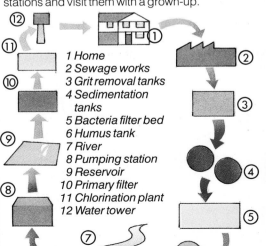

1 Home
2 Sewage works
3 Grit removal tanks
4 Sedimentation tanks
5 Bacteria filter bed
6 Humus tank
7 River
8 Pumping station
9 Reservoir
10 Primary filter
11 Chlorination plant
12 Water tower

EXPERIMENT·EXPERI

All living things need and use water in one way or another. Humans simply have to turn to a tap, collect some water in a glass and drink from it. But have you ever wondered how plants 'drink' their water supply? The first experiment on this page should give you some clues.

Experiment 3

Plants drink too!

In this experiment you can see how water gets in and out of plants by two very simple processes called *osmosis* and *transpiration*. You will need a carrot, a leafy twig, some sugar, a cork and plastic straw, two small jam jars, a large glass bowl, a wax crayon, some cooking oil and, or course, some water.

step 1

Make a deep hole in the center of your carrot and fill it with sugar solution. Make a hole in the middle of the cork and push the plastic straw through it. Use this to block the top of the carrot and seal the edges with melted wax. The put your carrot into a jar of water and leave it for several hours.

step 2

Now take your leafy twig and place it in the second jar of water. Pour a little cooking oil onto the water surface — this will act as a sealant to stop any water evaporating. Now mark the level of the water on the side of the jar with a wax crayon. Place the bowl over the jam jar and twig and leave them for a few hours.

step 3

When you return to your carrot, you'll see that water has risen up inside the straw. This happens because the carrot has absorbed the water in the jar by a process called *osmosis*. Osmosis happens because water always moves from a weak solution to a strong one. And because the liquid in the carrot cells is much stronger than the water in the jar, so the water molecules move into the carrot. Eventually some of this water passes into the sugar solution and forces it up into the straw.

When you return to your leafy twig, you should find that lots of tiny drops of water have collected on the inside of the glass bowl, and also that the water level in the jar has dropped slightly. This is because the water traveled up the twig until it reached the leaves. There the water vapor was given off through hundreds of tiny pores on the underside of the leaves. This is called *transpiration*.

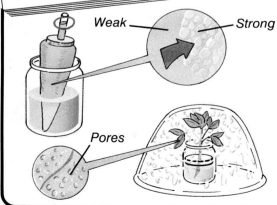

Weak — Strong

Pores

Experiment 4

Making a water turbine

With just two liquid soap bottles and a long knitting needle (or length of stiff wire) you can make your own water turbine. Simply follow the steps on this page. You will also need some scissors and a tube of model-making cement.

Step 1

Rinse out both of the bottles. Cut off the top and bottom of one of them so you are left with an open-ended cylinder. When opened out, this should give you a rectangular shape which can easily be cut into four equal strips.

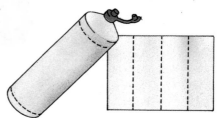

Step 2

Fold the strips in half and then stick one side of each to your second bottle, as shown. These will form the blades of your turbine and they should be set an equal distance apart. Leave until the cement has dried.

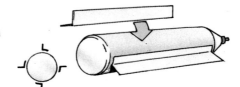

Step 3

Bend out the blades of your turbine so they stick out at right angles from the bottle. Then thread the needle in through the bottom of the bottle and out through the top. You should now be able to spin your turbine freely round the needle while holding both ends firmly.

Now take your completed turbine to the bathroom and hold it under a running bath tap. Notice what happens. The flow of water will move your turbine round and round, just as it does with a big water wheel or hydro-electric turbine. In both watermills and dams alike, this movement is harnessed to create a useable force.

WATER IN DISGUISE

Solid water is useful stuff. We pack fresh foods in it to stop them going off, we put it on swellings to make them go down, we skate on it, and we even put solid water into liquid water to make our drinks cold and refreshing. Of course, solid water, as you'll probably have guessed, is better known as ice.

Ice is just one form of water in disguise and, apart from being very useful in many ways, ice can be dangerous, too. In 1912 the huge ocean liner *Titanic* set off on her maiden voyage from Britain to North America. In mid-Atlantic she sailed into thick fog, hit an iceberg, and sank with the loss of hundreds of lives.

One of the strange things about ice is that it floats. Good scientists know that when a substance is cooled it contracts (shrinks). It becomes more dense — which means that the same volume of a substance weighs more when cold than when warm. Now the iceberg, being solid and colder than the water around it, should be heavier too, and should sink to the bottom of the sea. So why is it that icebergs and ice cubes float?

The answer lies in the fact that water acquires some peculiar properties when it turns from liquid to solid. If you had lots of jars full of water, each one the same size but at a different temperature, then they would all weigh different amounts. Warm water would weigh less than cold water, for instance. But the water at 39.2°F (4°C), would be the heaviest. This is because water shrinks as you cool it to 39.2°F, but then as it gets colder it starts to expand again! Even when it goes below 32°F (0°C) and turns to ice it still expands — which is why ice is lighter than water and can float

in it. Apart from ice, water also appears in another form — steam. Steam is a mixture of water vapor, which you can't see, and tiny water droplets, which you can. The water vapor is invisible, and just as water turns solid at 32°F (0°C), so it turns to steam at 212°F (100°C). In a solid substance all the molecules are arranged in neat criss-cross rows and can't move. In a liquid substance the molecules are able to move around, but they are still

*Frozen water here takes the shape of icicles,
formed as water drips from the granite rock
and freezes into the cold winter air.*

Photo: Science Photo Library / M Dohrn

HOT OR COLD?

You may find it odd that water freezes at exactly 0°C and boils at exactly, 100°C. These two numbers must seem extremely convenient! Of course what happened was that the centigrade temperature scale was arranged around the freezing and boiling points of water, since these properties are so important. Water's freezing point was simply called 0 and its boiling point on a Centigrade scale (centi = 100, grade = divisions) was called 100.

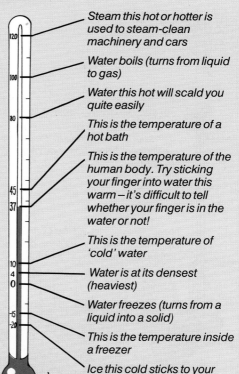

Steam this hot or hotter is used to steam-clean machinery and cars

Water boils (turns from liquid to gas)

Water this hot will scald you quite easily

This is the temperature of a hot bath

This is the temperature of the human body. Try sticking your finger into water this warm – it's difficult to tell whether your finger is in the water or not!

This is the temperature of 'cold' water

Water is at its densest (heaviest)

Water freezes (turns from a liquid into a solid)

This is the temperature inside a freezer

Ice this cold sticks to your fingers and 'burns'

If you have a 0-100°C thermometer you might try measuring the temperature of water around the house. How much colder does a glass of water from the cold tap get if it's left in the fridge for half and hour, or in the sun for 10 minutes? Try measuring the temperature of your bath, too. Is 45°C too hot for you?

quite close together. In a gas the molecules suddenly become hundreds of times farther apart, and much faster-moving as well. It is this rearrangement of its molecules that enables water to appear in its three different guises. Look at the steam rising from the kettle next time it boils, the ice forming on the inside of the freezer and the water drip-dripping from the tap. Strange, isn't it, that they're all really the same thing!

ICE AND STEAM

Ice and steam are not used as much in industry these days as they were in the last century. Before fridges and freezers were invented many foods were packed in ice to keep them fresh. Fish caught at sea were packed in ice in the ship's hold, and fresh meat being transported by trucks or trains was kept cold in the same way. The low temperature stopped any bacteria from breeding too fast and making the food go bad. Even lettuces and other vegetables were packed in ice, to keep them crispy and fresh-tasting. Nowadays we use refrigerated trucks, freezer containers and cold stores to keep food fresh by turning the water actually inside the food into ice.

Steam power is also used less these days. During the time of the Industrial Revolution virtually every piece of machinery was powered by steam. Cars, railways locomotives, and even sewing machines were worked by steam engines. Then along came the gasoline engine, and the age of steam died.

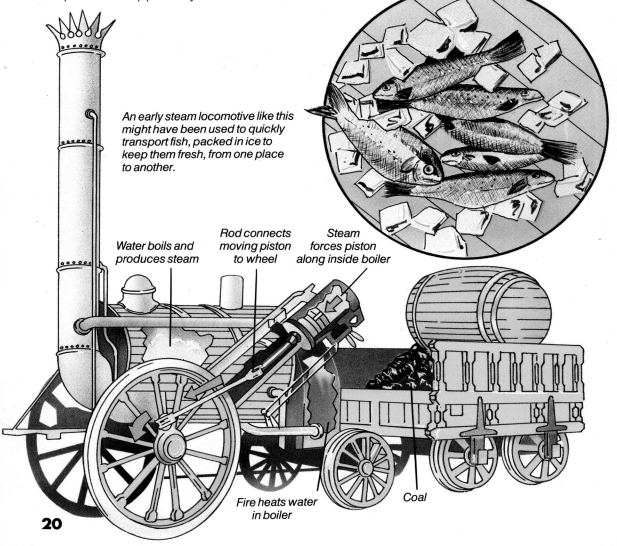

An early steam locomotive like this might have been used to quickly transport fish, packed in ice to keep them fresh, from one place to another.

Water boils and produces steam

Rod connects moving piston to wheel

Steam forces piston along inside boiler

Fire heats water in boiler

Coal

Science in action

Frozen fish?

In very cold weather you will notice that ice forms on the surface of most ponds. But even in a really hard winter the water near the bottom never seems to freeze. The reason for this is to do with the density of water (see page 18). As you know from the thermometer readings on page 19, water is at its heaviest at 4°C, so it sinks to the bottom of the pond. The colder water floats on top, and turns to ice when it reaches 0°C. Ice (surprisingly) is a good insulator and acts like a blanket to stop the water near the bottom from freezing. So the pond creatures can swim around near the bottom in the water at 4°C. This may seem cold, but it's better than being frozen to death!

Tea's made!

When a kettle whistles, the noise is made by the water boiling and giving off steam. This steam rapidly expands, rushes out of the kettle past a small hole and causes the noise. It's just the same as you blowing air through your pursed lips and making a whistling noise.

Ice burst the pipes

If water in a pipe gets to 0°C or below, it freezes. And when water freezes it also expands (as you can find out on page 22), and expanding ice is amazingly powerful. It can easily split the pipe and push apart any joints in the piping. When the weather warms up the ice melts and the leak becomes all too obvious. Moral: make sure all the water pipes in your house are well insulated!

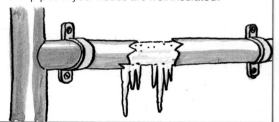

Science discovery

James Watt and the steam engine

James Watt is often said to have invented the steam engine by watching steam coming out of the spout of a kettle. However, he didn't actually invent the steam engine, but he did make a lot of improvements to its design. The first steam pump (used to get water out of coal mines) was invented by Thomas Savery in 1698. Around 1770-80 James Watt designed a much-improved version that converted the to-and-fro movement of the steam piston to a round-and-round rotary movement useful for powering

other machinery. And by 1782 Watt's engine was being used to power machines in factories all over the country. In fact, it was the invention of the steam engine that began the Industrial Revolution when people began to be replaced by machines that could do their work.

On the previous pages you can read about water in its two most unrecognizable states – ice and steam. There are many experiments that you can do to see for yourself some of the properties of water in its other forms. You'll find two on this page. Remember that steam is very hot and will scald you, so be careful in Experiment 6.

Experiment 5

The great ice experiment

You have read on the previous page about some of ice's strange properties. Now's your chance to see them in action! You will need a small, screw-top glass bottle full of water, some ink, an ice tray and a glass of water. You'll also need the use of a freezer or the freezer compartment of a fridge.

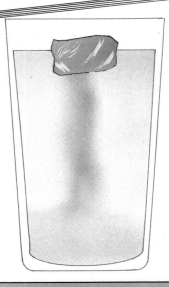

Step 1

Make sure the bottle is full of water and that the lid is screwed firmly on. Then put it into the freezer compartment of your fridge (let your parents know what you are doing). Stand it on a small tray to make clearing away simpler. Leave the bottle overnight and when you return you will find it has broken. This happens because, as the water froze, it expanded and forced the sides of the bottle, breaking the glass. This should give you an idea of how powerful ice is, and why water pipes are always in danger of cracking when winter is upon us.

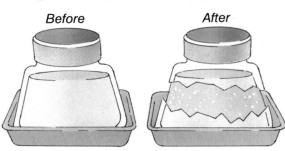

Before   *After*

Step 2

Dilute a small amount of ink (in a color of your choice) and fill one or two sections of your ice tray with the colored solution. Put it into the freezer compartment and leave overnight or until frozen.

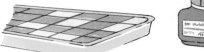

Step 3

Once frozen, remove a cube and place it in your glass of water. Watch what happens as it melts. Because water's density varies with temperature (as you can read on page 18) you will see that the colored water sinks to the bottom of the glass. What do you think would happen if you put another colored cube in hot water? Try it and see.

Experiment 6

Full steam ahead!

Make this simple model to see how steam can be used in a practical way. You will need a metal cigar tube, a cork to fit the tube's end, some strong wire, two small candles, two large nails, a piece of balsa wood and a box of matches. A pair of pliers and a small saw will also come in handy.

Step 1

Check that the cork makes an air-tight fit in the tube, then pierce a small hole through it with a nail or similar object. Next, take two equal lengths of wire and wrap them around the cigar tube as shown, twisting the wire together with a pair of pliers. Make sure that at least 6 in (15cms) remains at each end.

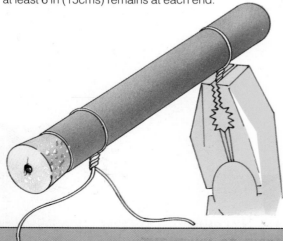

Step 2

Cut your balsa wood into a boat shape. Bang a large nail in at either end to act as a keel. Now turn the boat the right way up and fix your two small candles side-by-side as shown. It would be best to secure them with some of their own melted wax as they will then be easy to remove and replace.

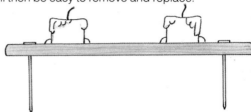

Step 3

Wrap the cigar's wire around the wood so the tube rests just above both candlewicks. Then fill the tube a quarter full with water and replace the cork. Now your boat is ready to sail! Place it in a bath tub, or perhaps a pond if the weather's calm, and carefully light the candles. It might be wise to get one of your parents to light the matches – they can then be present for the launch too! As the water heats up, so steam will form inside the tube. The steam expands and is forced out at high pressure through the only means of escape – the hole in the cork. This will be enough to drive your boat along. How far can your boat travel on one tank? Would it move faster if you added a third candle?

WET OR DRY?

People are always talking about the weather, and water is always there in some form. Think of the clouds on the horizon, the rain beating on the leaves, the hail clattering on the roof, the snow drifting gently, the frost glistening on the hedge, and the fog and mist obscuring the motorway ahead. These are all different forms of water in the weather — it appears in its different guises, as a solid, a liquid or a vapor, depending on the size of the water droplets and the temperature of the air around them.

Even ordinary air contains water, as invisible water vapor. This gets into the air from all sorts of different sources — by evaporating from the seas and rivers, from vapor given off by the millions of plants, even from the sweat given off by our own bodies. But of course the atmosphere can't hold an unlimited amount of water vapor so it all ends up coming back down to earth again in one form or another. You can find out why and how on the next page.

The amount of water vapor in the air is called the *humidity*. When there is a lot we say it is humid. We feel sticky because the watery sweat from our bodies cannot evaporate into the air — for it is already full of water vapor. When there's only a little vapor in the air, we say it feels dry. Now our sweat can evaporate easily and pass into the air as water vapor.

How many different forms of water in the weather can you think of? Note them down in the *Science Notebook*. Some are shown on this page.

Here are three forms of water in the weather: as clouds in the sky above, as snow coating the bare branches and as rain in a fierce thunder storm.

Photos: Science Photo Library / R Wetmore & (top right) M Dohrn

ENDLESS JOURNEYS

Turn on a tap and out comes some water. It flows down the drain into the sewage system and to your local water works, where it is filtered and cleaned (as you can see on page 15). Then it flows into the river and makes its way to the sea. After drifting about for a bit it is evaporated from the surface by the heat of the Sun and rises into the atmosphere. Up there's it's cooler and the water vapor condenses into tiny droplets that make up a cloud. The cloud is blown over the hills, rises higher to where it is even cooler, and the tiny droplets condense further into bigger drops that fall as rain. The rain soaks into the earth and works its way back into the rivers again. Some even lands in your local reservoir and flows into the water system. It is purified and fed into the water main that leads to your house. And, lo and behold, a month later, it comes out of the tap again!

Water on the move

Of course this is not strictly a true story, but it does show that water is always on the move. The reason why your tap does not run out of water and that rivers never cease to flow is that water is being constantly cycled, from sea to atmosphere to rain to soil to river and so on, in an endless variety of journeys. The Earth today contains as much water (in all its various forms) as it did when it was formed millions of years ago. And the reason for this is simply that water carries on going round and round. You can probably work out some journeys on the cycle for yourself. On a really slow route the water would get stuck deep in the ground or the sea for thousands of years. What is the fastest cycle you can think of? And where does the power for the cycle come from? Thinking of water being evaporated by the heat of the Sun should give you the answer.

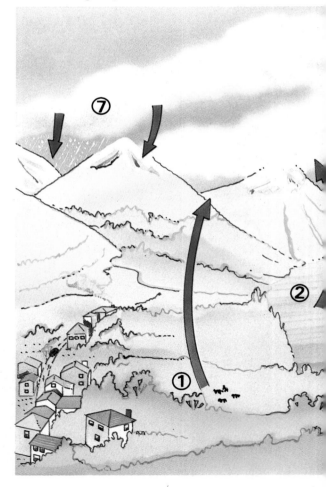

Key to water cycle diagram

1. *Animals and people breathe out water vapor into the air.*
2. *Water evaporates from the soil and passes into the air as vapor.*
3. *Plants give off water vapor through their many leaves.*
4. *The Sun's heat evaporates water from lakes and rivers.*
5. *More water evaporates from the sea and rises high into the air. There the vapor condenses into tiny droplets to form clouds.*
6. *Sometimes the droplets condense further and fall as rain over the sea.*
7. *The coulds are often blown over hills where the air is cooler. Here the droplets may freeze and fall to the ground as snow.*

Science factfile

Where's all the water?

▶ At any one time, the vast majority of the Earth's water is in the seas and oceans – 97.1% in fact. The other 2.9% is divided up between lakes and rivers, ice caps and glaciers, streams and ponds, the soil and the atmosphere.

▶ The wettest place on earth is reckoned to be Mt. Wai-'ale-'ale in Hawaii where it rains nearly every day!

▶ The driest place must surely be the Atacama Desert in Chile where it didn't rain at all for 400 years, until 1971!

▶ The greatest rainfall ever recorded was in the Indian Ocean in 1952 when over 71 inches (1,800 mm) came back to earth in 24 hours.

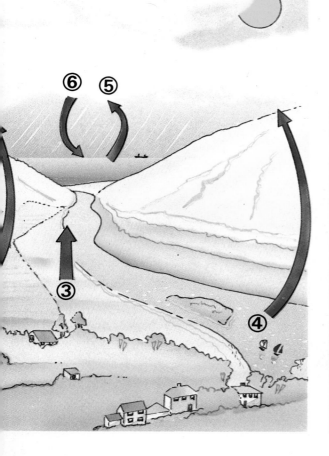

Science in action

Will it rain?

All clouds are made of tiny water droplets floating along in the air. But only certain kinds of cloud contain droplets that are big enough to fall as rain. You can often identify these rain clouds by their shape and color. Try looking at the sky first thing in the morning and predicting whether or not it's going to rain from the type of cloud you see.

Cirrus clouds are high, white and wispy. They don't bring rain.

Cumulus clouds are fairly high, white and puffy. They look a bit like cotton wool and they don't bring rain.

Nimbus clouds are low, grey and thick. They do bring rain.

Cumulonimbus clouds are low, grey and towering. They usually bring rain and also thunderstorms.

Water plays an enormous part in our weather, and because it is always present you can actually measure the amount of it floating about in the air and use this as a guide to the day's weather. A simple instrument called a *hygrometer* will do this. Of course, professional weathermen don't use only this method nowadays. They have satellites stationed high above the ground to monitor the world's cloud movements, and also computers to help them predict what weather is to come. But if you want a simple and effective weather predictor, here's your answer!

Experiment 1

The whirling weather wheel

To make your own hygrometer you will need a round tin can about ⅜in (1cm) deep and 2¾in (7cm) in diameter, a small block of wood, a needle, a nail, a bead, a cork, some writing paper and a strip of photographic paper. A file, a pair of scissors, a pen, a hammer and some glue will come in useful too.

Step 1

Punch holes through the middle of the base and lid of your can. Then make further holes in the base to allow a passage of air in and out. Cut out a circle of writing paper the size of the can's lid, an arrow just under the length of its diameter, and a small strip about ⅜in (1cm) wide by 5½in (14cm) long. Put these aside.

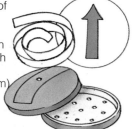

Step 2

Cut a strip of photographic paper about 6in (15cms) long and ⅜in (1cm) wide. Coat your strip of writing paper with glue and wrap it round the center of the needle. Leave a small piece of the glued paper showing and attach your photographic paper to this. Then wind it round into a coil, as shown.

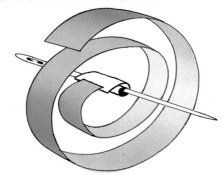

Step 3

Nail the base of your can to the block of wood to form a stand. Then push one end of the needle through the central hole. Let the photographic paper uncurl naturally and stick its free end to the inside of the can.

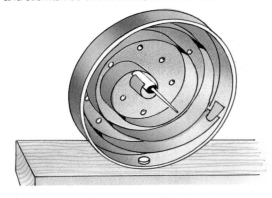

Step 4

Glue your circle of writing paper to the front of the lid. Then put the lid and base together so the needle is pointing out through the paper. Thread the bead onto the needle, then your paper arrow and, finally, the cork. Make sure that the arrow is gripped firmly by the bead and cork so that if the needle is turned, the arrow turns as well.

Step 5

Putting the readings onto the face of your hygrometer is the next step. You'll find that the paper pointer moves around depending on the state of the weather. This happens because when the air is heavy with water (and it's likely to rain), the photographic paper absorbs the moisture, expands, and consequently unwinds inside the tin, causing the needle to move as well. When the air is dry (and it is likely to be a fine day) the paper will dry out and shrink again. On our hygrometer the pointer will move to the right when wet, and to the left when dry. Rather than waiting for the weather to change, you could try marking up your model in obvious conditions. For instance, if you're having a bath and the room is full of steam, you will know that where the pointer sits is likely to be the wettest point on your hygrometer. An airing cupboard may be the other extreme.

Nature's weather predictors

Hygrometers, weather satellites, computers and lots of other things besides help us to forecast the weather. But did you know that Mother Nature has her very own weather predictors too!

A pine cone is a very good, and extremely simple, example. When a cone is open the weather is likely to be dry, but when the cone is shut it's likely to be wet. The reason for this is that if the

cone dropped its seeds in wet weather they would be too waterlogged to blow away in the wind. So the cone doesn't open until the weather is dry and its seeds have a chance of traveling to new plates in which to grow.

Seaweed is another natural weather predictor, and an even simpler one to understand. When the air is damp, the seaweed absorbs the moisture and becomes wet and slimy. But when the air dries up, the seaweed's water evaporates and leaves the seaweed brittle and dry!

SINK OR SWIM?

You'd be quite surprised if you threw a piece of wood into water – and it sank. Yet there is a type of wood that's so heavy it sinks in water, even when freshly cut and dried. It's called ebony, it's black in color, and at one time it was used to make the black keys on pianos.

You'd also be surprised if you threw a lump of metal into water – and it floated. Yet there are metals that are so light they'll float, though most of them (like lithium and sodium) are chemically unstable and react quite violently with water. Such metals are often mixed with other, heavier metals to make the light but strong alloys that are used in space vehicles, racing cars and so on.

So how can you tell whether something will float without actually putting it in water? One way is to measure its *density*. You can find this out by dividing the weight of the object (in grams) by the volume (in cc's) to give you the density. If the answer is less than one, the object will float. But if the answer is more than one, it will sink. Why the magic number of one, you may well ask? And the answer, of course, is because the density of water is one! This may seem remarkably convenient but, like the temperature scale on page 19, the density scale is based on water since water is so abundant and important.

However, not all water is the same. The density of pure water is one, but what about the density of sea water which has lots of substances dissolved in it? You can test this yourself by dissolving some salt in a measuring jug of water. You'll find that as you add the salt, the weight

PLIMSOLL LINES
The mark of the side of the boat in the picture below is called a Plimsoll line and it shows the ship's captain how low down in the water the ship is sitting. You can see it is made up of several marks. This is because a ship floats at different levels depending on the kind of water it is in.

of the solution goes up – but its volume stays the same. This means the density of the solution is going up as well!

Density is very important because it determines how easily objects float or sink; the more dense the liquid, the more

Photo: M Little

weight it will support. Our bodies must be a little more dense than water since we tend to sink if we don't swim. In the fresh water of a swimming pool, for example, we have to put quite a lot of effort into staying afloat, while in the sea, where the water is denser, it is slightly easier. But in the Dead Sea, in Israel, it is easier still. This land-locked sea has very salty water indeed, which is very dense and, therefore, very easy to float in. It must be very difficult indeed to drown in the Dead Sea!

FLOATING ABOUT

Before the early 1800's ships were made of material that floats, such as wood or animal skins stretched over poles. But with the coming of steamships people wanted bigger and stronger boats to carry more passengers and cargo. The engineers tried to work out how they could make a ship of iron that would still float. Many non-scientists laughed at the idea: iron was much denser than water, so how could an iron ship possibly float? As soon as it was launched it would sink straight to the bottom!

But the engineers knew that whether something floats or sinks depends not only on its density, but also on its shape. The shape of an object determines how much water it displaces – that is, how much water is 'pushed aside' when the object is put into water. They remembered the principle of displacement discovered by the Greek mathematician, Archimedes, over 2,000 years ago (which you can read about in the panel).

What is displacement?

The principle of displacement works as follows. First of all, weigh something, such as a block of metal. Then put it in some water and weigh it again. You will find that it seems to weigh less. This is because the water is actually helping to support the object to a certain extent. If you push a floatable object,

1 Here is a tank of water. Some things will float in it, others will sink. (And, of course, our friends in the picture can do either.)

2 A solid block of iron displaces (pushes aside) some water, and the level in the tank rises. The displaced water weighs much less than the block, though, and the block sinks.

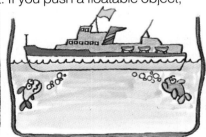

3 The same amount of iron is made into a ship shape. This displaces far more water, as you can see because the level rises much more. The weight of the displaced water equals the weight of the ship, and the ship floats.

Science discovery

Archimedes and the crown

There is a famous story about how the mathematician Archimedes discovered the principle of displacement. The local king, Hieron, had a gold crown made, but he suspected that the goldsmith had cheated him by mixing a less valuable metal with the gold. He asked Archimedes to solve the problem – but without damaging the crown.

One day Archimedes was getting into a full bath when he noticed that as he entered the water, some of it splashed onto the floor. The idea of displacement came to him, and he jumped out of the tub and ran naked through the streets shouting "Eureka" (I have found it!)

Archimedes' problem was that he knew how much the crown weighed, but he couldn't measure its volume since it was such an awkward shape. So he made a block of gold of the same weight as the crown. He then put the block and the crown into water and measured how much water each one displaced. It turned out that the crown displaced more water than the gold block, so the crown could not have been made of pure gold but of gold mixed with a less dense, and less valuable metal. Archimedes was rewarded by the King, and the goldsmith was jailed for his deception.

such as a piece of wood, down into some water you'll be able to feel this force trying to push it up again.

So how much less does the object weigh? Archimedes discovered that the object's loss of weight is the same as the weight of water it displaces. You can see displacement for yourself by simply putting some water in a glass and marking the level with a crayon. Now drop an object into the water and you will see that the level rises. It is the weight of this 'pushed aside' (or displaced) water that counts, for if it weighs more than the object itself then the object will float, and if it weighs less then the object will sink.

Now, the amount of water displaced by an object has a lot to do with its shape. If the object is a solid block it won't displace much water. But if it is made into a cup or bowl shape it displaces much more (provided the material is watertight, of course!) In fact, when a bowl shape is floating it is not only the bowl doing the displacing but also the air in it – and the air weighs almost nothing.

Will it float?

This was the reasoning the steamship engineers went through. They were sure an iron ship could float provided they worked out the weight of the finished ship and then made sure the shape was big enough to displace more water than this. They were right. In 1823 the first iron ship, the *Aaron Manby,* was launched and floated perfectly.

Science in action

The rise and fall of the sub!

The submarine is one type of boat that relies on the displacement principle not only to keep it afloat but to help it sink as well! Here we see how a sub manages to float, then sink, then return to the surface.

1 The sub floats like any other boat, by weighing less than the amount of water it displaces. But there is one difference – the sub sits quite low in the water, and is almost ready to sink.

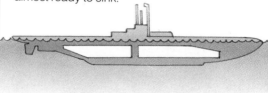

2 To dive, the sub allows sea water into its ballast tanks. It now weighs more than the water it displaces, so it sinks.

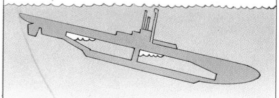

3 To resurface, the sub pumps air from its compressed-air tanks into its ballast tanks, so-forcing the sea water out again. The sub is now lighter than the amount of water it displaces, so back it goes to the surface. (The air inside the ballast tanks weighs as much as it did when in the compressed-air tanks, of course. But it now takes up room previously occupied by much heavier water.)

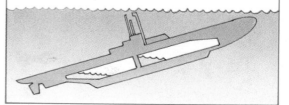

On this spread are two experiments to do with floating and sinking. If you would like to test Archimedes' theory of displacement as well, you could try this very simple test. Get a large bowl full of water and a ball of clay. If you try putting the clay into the water, it will sink. But after reading pages 32 and 33 can you think of a way to make it float? Remember, clay can be moulded into any shape.

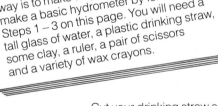

Experiment 8

Making a hydrometer

On page 30 you can find out how to measure the density of a liquid by dividing its weight by its volume. But a far easier way is to make a hydrometer. You can make a basic hydrometer by following Steps 1 – 3 on this page. You will need a tall glass of water, a plastic drinking straw, some clay, a ruler, a pair of scissors and a variety of wax crayons.

Step 1
Cut your drinking straw so you are left with a piece 5in (about 12cms) long. Stick a lump of clay, about as big as a pea, on one end of the shortened straw.

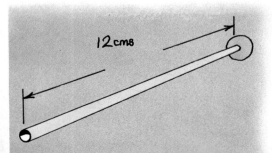

12 cms

Step 2
Put the straw into water about 4in (10cms) deep. If the straw sinks to the bottom of the glass, you have too much clay; if it floats but tilts to one side, you have too little clay. Keep adjusting the amount of clay until the straw floats upright in the water.

Step 3
Put a depth scale on your straw by lying it next to a ruler and marking off ⅜in (1cm) lengths with your colored crayons. Remember to press firmly so your scale will be quite clear, but be sure not to split the straw.

Step 4
Now mark a level on the side of your glass and fill it up to this point with water. Drop in your hydrometer and measure the amount of straw that sticks up above the surface. Then try testing other liquids for their density – the more dense the liquid, the higher the straw will float. Try comparing water with brine or milk. But if you want to test liquids like methylated spirits be sure to ask for permission first.

 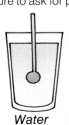 

Brine   Water   Meths

Experiment 9

Operation octopus!

Here's your chance to trick your friends by making a mini-octopus that dives up and down without even being touched. Before you start, make sure you have a plastic pen lid, a supply of paper-clips, two colored beads, some strands of wool, a lump of clay and a large, plastic screw-top bottle.

Step 1

Make a small hole through the bottom of the pen lid. Now you can string two paper-clips onto the lid as shown, hooking the first one through the hole and linking the second one up to make a chain.

Step 2

Cut four strands of wool, each measuring 5in (12cms). Fold them in half and stick the folded ends to the top of the lid with clay. This will give your octopus his eight legs. Now use the beads to give him some eyes.

Step 3

Fill your bottle with water, then drop in your octopus, making sure you keep him upright as he enters the water. He should sit at the surface – if he sinks then your paper – clips are too heavy. Try adjusting the weight until he just floats happily at the top. Screw on the lid of the bottle.

Step 4

Now, can you think of a way to make your octopus dive to the bottom of the bottle without actually touching him? No? Well, try squeezing the sides of the bottle and watch what happens. As you squeeze, down goes your octopus; stop squeezing and up he comes again! This happens because the octopus is kept afloat by a tiny bubble of air trapped in the top of the pen lid. When you squeeze the bottle, the water compresses this air bubble into a smaller space which means that your octopus contains more water and is, therefore, a tiny bit heavier. So he sinks to the bottom! When you stop squeezing, the air bubble expands again, pushes out the water and back goes your octopus to the surface. You can see from reading page 33 that this is just how a submarine makes itself sink to the bottom of the ocean and then float back to the surface – by changing the amount of air it contains. Now try your diving trick on your friends. Can any of them work out how to make the octopus dive?

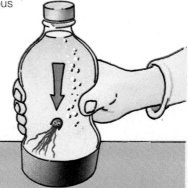

WALKING ON WATER

ater, like you, has a sort of 'skin'. But it's not like the skin that we have, nor like the skin that forms on top of hot drinks like coffee and cocoa. Nevertheless, small water creatures like the pond skater rely on water's skin to hold them up above the water itself. Watch them skating about on your local pond in summer — they can cross the surface without even getting their feet wet!

To understand how the surface skin of water comes about we have to think about molecules of water. If you had a glass full of water, for instance, the molecules in the middle of the water would each be surrounded by other molecules, and would be attracted equally to all of them. But a molecule at the surface would only have other molecules to the sides and below. The result of this is that the water molecules at the surface are attracted sideways and downwards, to the other water molecules, rather than upwards to the air. And it is this pulling together of the surface molecules that

creates the force—called surface tension—which forms water's so-called skin!

One easy way to see this tension for yourself is to fill a narrow glass full to the brim with water. Do this very slowly and you will see that, as the water reaches the rim, it begins to swell up above the top of the glass in a curve. It is the surface tension that is holding the water in and preventing it from spilling over.

The pond skater isn't the only creature to use surface tension. To the creatures in the pond beneath, the surface skin must be like a kind of ceiling. Mosquito larvae hang from this ceiling, poking their tails through into the air in order to breathe. A dragonfly climbing down a reed stem towards the surface has to fight very hard to break through the tension so it can lay its eggs in the water below. In fact, the smaller you are, the harder it is to break the tension and microscopic animals can easily become trapped in the skin if they swim too near the surface.

THE FORCE AT THE SURFACE

You can see from this diagram how water molecules at the surface are only attracted downwards and sideways to their fellow molecules, rather than upwards to the air.

Air   Water molecules   Strong links

Photo: Gower Scientific Photos

WETTING & WASHING

Do you know how to make water wetter? Try washing your hands with cold water on its own. The water tends to roll off in drops and doesn't wet you very much. This is because your skin produces an oily substance that repels water – and does not break the water's surface tension.

Making water wetter

Now wash your hands with soap and water. Immediately you should notice the difference. The water seems to be wetter. This is because the soap acts as a 'wetting agent' – its molecules get in between the water molecules and destroy their attraction to each other. This reduces the water's surface tension, so that its skin is weaker and more easily broken. Soap and water together wet your hands, and dissolve the grease that traps dirt on your skin, and in this way they help to wash you clean. Many 'wetting agents' have been made that work much better than soap. They are used in special cleaning processes where every bit of dirt has to be wetted and washed away. You can see for yourself how soap breaks water's surface tension by doing the experiment on page 41 of this book.

Keeping water out

The opposite of a wetting agent is a waterproofing substance that repels water and does not break the surface skin. These substances usually have an oily, waxy or greasy feel. Many aquatic creatures have fur or feathers that are coated with a waterproofing substance. If you find a seagull's feather, for example, you'll notice that it is almost impossible to make it wet. The water runs off it in large drops, due to the waterproofing oils that coat the vanes of the feather. You can see some more creatures and their methods of keeping dry in the panel opposite.

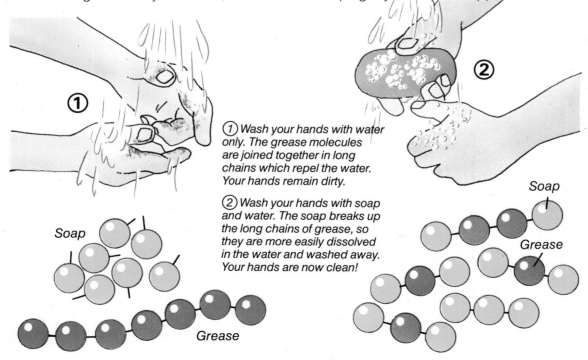

① Wash your hands with water only. The grease molecules are joined together in long chains which repel the water. Your hands remain dirty.

② Wash your hands with soap and water. The soap breaks up the long chains of grease, so they are more easily dissolved in the water and washed away. Your hands are now clean!

Soap

Grease

Soap

Grease

Science in action

An otter wears two coats. The outer one is made of long, coarse guard hairs for color camouflage. The fur of the inner coat is waterproof, so the otter's skin does not get wet.

Many of the creatures that live in or near water have their own special methods of keeping dry. Many have waterproofing oils to prevent their fur or feathers from becoming waterlogged in the first place while others, like the cormorant, are not so lucky. They have to work hard drying off their wet bodies in the wind and sun.

A duck preens itself. It is using its beak to spread body oils over its feathers from special glands above its tail. This keeps the feathers waterproof.

A cormorant has to dry its wing feathers after a swim. This bird does not have a very good waterproofing mechanism (unlike most other water birds), and can become so waterlogged that it can't fly! This is one reason why you'll often see it drying off like this on the cliff tops.

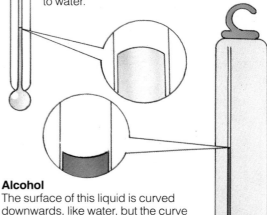

Science project

Looking at the surface skin

Water is not the only liquid that has a surface skin. Most other liquids have a skin too, though they are each different in nature. Have a look at other liquids in narrow glass tubes, where the skin is most obvious. Can you see the differences?

Mercury
This substance is really a metal, but nevertheless it takes the form of a liquid at normal room temperature. Look closely at its surface in a mercury thermometer (the maximum/minimum type is best). It is curved upwards, the opposite way to water.

Water
In a fine-bore glass tube you can clearly see the shape of water's skin (this is called the *meniscus*). It curves downwards.

Alcohol
The surface of this liquid is curved downwards, like water, but the curve is shallower. You can see this in a glass-tubed freezer thermometer, which usually has red-dyed alcohol inside it to measure low temperatures.

On this page are several fun ways of experimenting with surface tension. Remember, though, that there are lots of simple things that you can do to prove its existence as well – like filling a small glass to the brim (see page 37), or simply watching a pond skater walking or standing on the surface of a pond, or even watching out for dust trapped in the surface film.

Step 1

Put a small square of tissue onto the fork and lay the needle on top. Lower the fork into the dish of water so the tissue paper and needle are left sitting on the water's surface. Leave for a few minutes.

Step 2

Put your cork disc on top of a quarter or fifty-cent piece but position it to one side, as in the picture. Now drip water onto the coin's surface using the pipette. Watch carefully what happens to the cork as you add the water drops.

Step 3

Look back at the bowl of water. You'll see the tissue has sunk to the bottom while the needle is still sitting on the surface. This happens because the tissue soaks up the water and sinks without disturbing the surface tension, leaving the needle to be held up by water's mysterious 'skin'.

Now look at your cork disc. You may have noticed that, as you added the drops of water, the disc moved to the exact center of the coin. This is because, as you add more drops, the tension of the water at the edges of the coin increases while it remains constant in the middle. The disc is pulled to this region of lower tension which is why it suddenly moves as if possessed by magic! Ask your friends if they can find a way of moving the disc to the center without actually touching it.

Experiment 10

Seeing is believing!

Just follow the steps below to prove to yourself that water has a 'skin'. All you need is a small dish full of water, some tissues, a needle, a fork, a cork disc (cut the end off an ordinary cork for this), a pipette or eye-dropper and a large coin.

Experiment 11

The water waltzers

In this experiment you can see for yourself how surface tension can be used to make things move along in the water. You will need the following: a cork disc and four small cork pieces, four bits of soap, some glue, a bowl and four needles. To make the dancing models you will need some stiff cardboard, colored pens and some scissors.

Step 1

If you like drawing then you can design your own dancing figures. If not, then trace the ones here and transfer them onto your cardboard. Go over the figures with your colored pens and then cut them out. Remember to leave a flap at the bottom, like the one in the picture.

Step 2

Push one needle into each side of the cork disc, as in the picture. Then cut a wedge in each of the four remaining cork pieces and fill each wedge with soap. Add one bit of cork and soap to each needle.

Step 3

Stick your dancers to the disc by gluing the flap firmly to the cork. Now you are ready to start them dancing – just by putting them in some water! But first, a word of warning. Your water must be very clean and still (if there are any soap particles in the water then the surface tension will automatically be weakened). So wash and rinse your bowl well before filling it, then leave it to settle for a few minutes. Now float your dancers in the water and watch what happens.

Step 4

You will see that they begin to waltz round and round, as if pulled by some mysterious force. What is, in fact, happening is that the pieces of soap are stopping the water molecules at the surface from clinging together in the normal way. This weakens the tension behind the bits of cork, leaving them to be pulled forward by the stronger tension in front. The result – your dancers start waltzing round the bowl!

You can make a less elaborate model to show the same thing. Just cut a small boat shape from a piece of stiff cardboard, make a slit in the back and slip in a small piece of soap. Put this mini-speedboat into a bowl of water (making sure it's clean and still, as before) and whoooosh... off it goes!

THINGS TO REMEMBER

Here are some explanations of words in this book that you may find unfamiliar. There aren't the exact scientific definitions, because many of these are extremely complicated, but the descriptions will help you to understand the STEP INTO SCIENCE books.

ATOM The smallest particle of any single substance, like oxygen or iron. Two or more atoms joined together make a molecule.

CONDENSE To turn from a gas or vapor into a liquid.

DENSITY The measure of how heavy a substance is compared to how big it is. Lead has a high density, balsa wood has a low density.

DISPLACEMENT The amount of fluid 'pushed aside' by an object floating or immersed in it.

ELECTROLYSIS The breaking up of a solute by passing an electric current through it.

EVAPORATE To turn from a liquid into a gas or vapor.

FLUID A liquid or a gas (not a solid).

HUMIDITY The measure of how much water (as vapor) there is in the air.

HYDROMETER An instrument for measuring the density of a liquid.

HYGROMETER An instrument for measuring the humidity of the air.

ICE Solid (frozen) water.

MENISCUS The curve on the surface of a liquid caused by surface tension.

MOLECULE The smallest particle of a substance made up of two or more atoms. A water molecule is made of three atoms, two of hydrogen and one of oxygen.

OSMOSIS The tendency of water to move from a weak solution to a strong one, to make the two solutions of equal strength, through some sort of barrier or membrane.

PLIMSOLL LINE The marks on the side of a ship, at the water line, that show how low in the water the ship is floating.

SATURATED SOLUTION A solution that has the maximum amount of solute dissolved in it, so that any more solute cannot dissolve and stays as a solid.

SOLUTE The substance that is dissolved in a solvent to make a solution.

SOLUTION The result of one substance being dissolved in another. The substance which is dissolved is usually a solid and is called the solute. The substance it is dissolved in is usually a liquid and is called the solvent. The molecules of the solute separate from each other and float about freely in the solvent. An example of a solution is salt (the solute) dissolved in water (the solvent).

SOLVENT The substance that is capable of having another substance (the solute) dissolved in it to make a solution.

STALACTITE A rock column that hangs from the roof of a cave or overhang, formed by mineral deposits left when water evaporates. (A good way to remember this name is stalaCtite = Ceiling.)

STALAGMITE A rock column that rises up from the ground, formed in the same way as a stalactite. (A good way to remember this name is stalaGmite = Ground.)

STEAM A mixture of tiny water droplets and water vapor formed when water boils.

SURFACE TENSION The 'skin' on the surface of a liquid, caused by the pulling together of the surface molecules.

TRANSPIRATION The evaporation and release of water (as water vapor) through the tiny pores in a plant's leaves.

WATER A normally liquid substance made of molecules containing one atom of oxygen and two atoms of hydrogen. Pure water has no color, no taste, no smell, turns to a solid at 0°C and a vapor at 100°C. Its density is 1 gram per cubic centimetre, and it is an extremely good solvent.

WATER VAPOR Gaseous water.

EXPERIMENT AND PROJECT INDEX

Full steam ahead 23
Looking at the surface skin 39
Making a hydrometer 34
Making a water turbine 17
Making a crystal columns 11
Now you see it; now you don't! 9
Operation octopus! 35
Plants drink too! 16
Seeing is believing! 40
Splitting up water 10
The great ice experiment 22
The water waltzers 41
The whirling weather wheel   28